A

NEW SYSTEM

OF

MEASURES, WEIGHTS, AND MONEY;

ENTITLED

THE LINN-BASE DECIMAL SYSTEM;

AND

DESIGNED FOR THE ADOPTION OF ALL CIVILIZED NATIONS,

AS

THE ONE COMMON SYSTEM.

BY

W. WILBERFORCE MANN.

NEW YORK:
UNIVERSITY PUBLISHING COMPANY,
155 AND 157 CROSBY STREET.
1871.

A NEW SYSTEM OF MEASURES, WEIGHTS, AND MONEY; ENTITLED THE LINN-BASE DECIMAL SYSTEM; AND DESIGNED FOR THE ADOPTION OF ALL CIVILIZED NATIONS, AS THE ONE COMMON SYSTEM.

A

NEW SYSTEM

OF

MEASURES, WEIGHTS, AND MONEY;

ENTITLED

THE LINN-BASE DECIMAL SYSTEM;

AND

DESIGNED FOR THE ADOPTION OF ALL CIVILIZED NATIONS,

AS

THE ONE COMMON SYSTEM.

BY

W. WILBERFORCE MANN.

NEW YORK:
UNIVERSITY PUBLISHING COMPANY,
155 AND 157 CROSBY STREET.
1871.

JOS. J. LITTLE,
ELECTROTYPER, STEREOTYPER, AND PRINTER,
NEW YORK.

THE

LINN-BASE DECIMAL SYSTEM

OF

MEASURES, WEIGHTS, AND MONEY.

THE base of the new system of measures and weights, here offered to the consideration of Science and Commerce, and to the statesmen and Governments of the World, is its unit of linear measure or measures of length.

The one hundred millionth ($100,000,000^{th}$) *part of the quadrant of the earth's meridian*, or the one hundred millionth part of the distance from the equator to the pole, is assumed as its unit of lineal measure.

The justly celebrated and admirable "Metrical System" of France has previously taken for its standard the quadrant of the earth's meridian; and it establishes the *mètre*—the ten millionth part of the quadrant—as its unit of lineal measure and base. The expediency of assuming the *one hundred millionth*, instead of the *ten millionth* part of the quadrant, as the unit of the new system, will, it is believed, become apparent in the sequel. This length of ten million *mètres*, as the distance from the equator to the pole, was deduced from the great trigonometrical measurement of the meridian from Dunkirk to Barcelona, in 1806–7. From comparison of English standards with a copy of the *mètre* in possession of the Royal Society, Capt. Kater found the length of the *mètre* to be 39.37079 of the English standard (Phil. Trans., 1818). Mr. Bailly found the length of the *mètre* to be 39.3696786 inches of the Royal Astronomical Society's scale (Mem. R. A. S., vol. ix. page 133); from which, by reducing to the imperial standard, by the data given in the same memoir, the true length of the *mètre* is 39.370091 inches of the imperial yard. It is *one tenth* of this length—3.9370091 *inches of the British imperial yard*—which has been fixed upon as the *unit of lineal measure and base* of the new system of decimal measures and weights here described, and from which all its calculations are deduced.

NOMENCLATURE.—The nomenclature of the system has been invented with special regard to conditions of acceptability by all nations. The fitness and utility that would make names, throughout the system, indicate the nature of the thing named, were also borne in mind. The six units are called—

Linn (Lat. *linea*, a line) unit of lineal measure, or of length.

Arr (Lat. *area*, surface, area) unit of square measure, or of area.

Soll (Lat. *solidus*, solid) unit of solid measure, or of volume.

Capp (Lat. *capacitas*, capacity) unit of liquid and dry measure, or of capacity.

Pondd (Lat. *pondo*, weight) unit of gravity, or of weights.

Monn (Lat. *moneta*, money, coin) unit of money.

The unmistakeable cosmopolitan character of these names seems to fit them for general adoption. Neither by form nor origin do they point to any particular modern country. They can neither gratify nor wound the *amour propre* of any existing people. They are short, being all monosyllabic, and of easy utterance, beginning and ending with different letters, so that there can be no possible confusion among them, of sound. Throughout all the denominations, they are invariable, having no plural termination. We say: 1 linn, 25 linn—1 pondd, 85 pondd. Moreover, these names, to the ear and to the mind, by sound, as by origin, are instantly suggestive of the thing signified. *Linn* suggests lineal measure, and gives its name to the system. Arr suggests measures of area. Soll suggests solids; capp, capacity; pondd, weights; and monn, money. It should be especially noted, also, that in all the languages of Europe, very nearly the same pronunciation must be popularly given to these names of the units. By the accepted rules of pronunciation, the sound of the words when uttered, cannot materially vary, wherever the letters of our language are used. But the final consonants must remain double, or grave variations of pronunciation will ensue.

The value of the several units is as follows:

LINN, the 100,000,000th part of the quadrant of the earth's meridian.		
ARR, an area, or superficies of	1	square linn.
SOLL, a volume, or bulk of	1	cubic linn.
CAPP, a vessel whose capacity is of..................	1	" "
PONDD, the weight of distilled water	1	" "
MONN, a silver coin of the value of 5 francs.		

The multiples and divisions or fractions of the units.

All the multiples and fractions of the several units are *decimal;* and the denominations are formed by prefixing to the names of the units Greek numerals, from one to ten, to designate the multiples, and Latin numerals, from one to six, to designate the fractions. But, in the formation of the denominations, the seventh and ninth numerals of the ascending series, or multiples, and the fifth of the descending series, do not appear. The numerals are—

DECA (Gr. δέκα, ten)	tenth	multiple	10,000,000,000
ENNA (Gr. ἐννέα, nine)	ninth	"	1000,000,000
OCTA (Gr. ὄκτω, eight)	eighth	"	100,000,000
HEPTA (Gr. ἑπτὰ, seven)	seventh	"	10,000,000
HEXA (Gr. ἕξ, six)	sixth	"	1000,000
PENTA (Gr. πέντε, five)	fifth	"	100,000
TETRA (Gr. τέτρα, four)	fourth	"	10,000
TRIA (Gr. τρία, three)	third	"	1000
DUA (Gr. δύο, two)	second	"	100
HENA (Gr. ἕνα, one)	first	"	10
Unit..			1
PRIMI (Lat. primi, first)	first	division	...........$\frac{1}{10}$
BINI (Lat. bini, two)	second	"	...........$\frac{1}{100}$
TERNI (Lat. terni, three)	third	"	...........$\frac{1}{1000}$
QUARTI (Lat. quarti, four)	fourth	"	...........$\frac{1}{10000}$
QUINI (Lat. quini, five)	fifth	"	...........$\frac{1}{100000}$
SENI (Lat. seni, six)	sixth	"	...........$\frac{1}{1000000}$

In this system, the numerals employed do not—as in the French "Metrical System"—indicate the products resulting from multiplication of the units, nor the fractions resulting from their division; but they indicate the number and order of the several multiplications and divisions. To illustrate: *myria*, in the "Metrical System," expresses the multiple 10,000. In this system, *tetra*, the equivalent of *myria*, indicates the fourth decimal multiplication of the unit, producing, as is seen in the above table, the same multiple, 10,000. In the descending series, or divisions of the units, *milli* of the "Metrical System" expresses the one thousandth part of the unit. In this system, *terni*, the equivalent of *milli*, indicates the third decimal division of the unit, descending, as may be seen above, to its one thousandth part. The advantage of this plan of nomenclature, avoiding, as it does, the use of long, doubly and trebly compounded numerals in the higher and lower denominations, will appear as we proceed.

But the numerals of the Linn-base System do really specify as distinctly, and, perhaps, more conveniently than those of the "Metrical System," the actual products and fractions resulting from the multiplications and divisions of the units, which form the denominations of the system. For, it will be noted in the above table that our numerals invariably specify the *number of ciphers* required, when placed after the unit, to express the fraction resulting from a division, and the product of a multiplication. Thus, *tria*, indicating the third multiplication of the unit, specifies also the number of ciphers required to express the product; to wit, three, composing the multiple 1000. *Seni*, of the other series, indicating the sixth decimal division of the unit, specifies also, in the manner described, as the result of that division, the fraction, the 1000,000th part of the unit, expressed by six ciphers. So, *dua* specifies the multiple 100; and *primi*, the first division, specifies, by one cipher, $\frac{1}{10}$ of the unit, as the fraction arrived at by the first division.

In fact, for the use of Science (and the attention of scientific men is particularly invited to this feature of our nomenclature), the numerals of the system are not merely natural numbers.

They are *logarithms*, as well. They are the numerical exponents of a ratio. They form a series of numbers in arithmetical progression, susceptible of indefinite extension, and answering to another series of numbers in geometrical progression, also susceptible of indefinite extension. They are the logarithms of a system of logarithms whose base is 10. Thus:

hena.	dua .	tria .	tetra .	penta .	hexa .	&c.	arithmetical progression.
0 .	1 .	2 .	3 .	4 .	5 .	6 . &c.	
1 .	10 .	100 .	1000 .	10000 .	100000 .	1000000 . &c.	geometrical progression.

Our numerals, treated as logarithms, offer to science unlimited power of facile expression in the use of the Linn-base System. Its units may be raised by geometrical progression to any required power, which will be instantly indicated by the appropriate logarithm. The application of the logarithms for the multiplication of the units is illustrated in the Synoptical Comparative Table hereto annexed. But the rule in question applies no less to the submultiples, or divisions of the units, than to their multiples. In the monetary department of this system, logarithms, applied to the very large amounts prevalent now-a-days in the budgets and treasury reports of all nations, afford facilities that can hardly fail to be appreciated. The sum of one hundred million Monn, which, in natural numbers, must be expressed by eight ciphers after unit, is announced under the Linn-base Decimal System, by either of three short forms. It may be written *octamonn*, or more shortly, $Monn^{8}$, or simply M^{8}. Decamon, or Monn^{10}, or M^{10}, signifies ten thousand million (10,000,000,000) Monn; Monn^{11}, or M^{11}, one hundred thousand million; the numeral logarithm annexed to Monn, always indicating the number of ciphers required after unit to express the sum in natural numbers.

DENOMINATIONS.

LINEAL, OR LONG MEASURE--UNIT LINN.

DENOMINATIONS.	VALUE LINN.	AMERICAN AND ENGLISH EQUIVALENTS.	FRENCH.
Octalinn....	100,000,000	6213.713857 British statute miles.	
Pentalinn ..	100,000	6.2137 miles.	myriamètre.
Tetralinn...	10,000	0.62 mile = 1093.613639 yards.	kilomètre.
Dualinn	100	10.936 British imperial yards.	decamètre.
Henalinn ...	10	1.0936 yards = 39.37 inches.	mètre.
LINN.......	1	3.9370091 inches of imperial yard.	decimètre.
Primilinn...	$\frac{1}{10}$	0.3937 inch.	centimètre.
Binilinn	$\frac{1}{100}$	0.039370091 inch.	millimètre.

10	binilinn	(bnL)	make	1	primilinn
10	primilinn	(pmL)	"	1	linn
10	linn	(L)	"	1	henalinn
10	henalinn	(hL)	"	1	dualinn
100	dualinn	(dL)	"	1	tetralinn
10	tetralinn	(tL)	"	1	pentalinn
1000	pentalinn	(pL)	"	1	octalinn (oL)

The henalinn, divided into linn and primilinn (tenths and hundredths), would serve in commerce as our yard.

The tetralinn and dualinn would be the common itinerary measure, as we now use the mile and yard. Perhaps, as distances in the United States are so great, it would prove more convenient there to take the pentalinn as the usual measure of itinerary distance. The dualinn, double dualinn, and demi-dualinn would be used—as are their equivalents in France—for surveyors' chains, each link having the length of two linn.

The octalinn is given for the rendering of immense astronomical distances and other scientific uses. For such purposes, it is far more convenient and appropriate than any of the infinite variety of leagues and miles in common use all over the world, or than the French myriamètre. Astronomical distances are so great, that they are quite inconceivable when expressed in miles; but the octalinn would promptly convey to the popular mind, and to intelligent youth, a definite and correct idea of many of them, for it will be remarked that the octalinn has precisely the length of the standard-base of this new system of measures. It is the quadrant of the earth's meridian—*one hundred million linn*—the distance from the equator to the pole—exactly one fourth of the circumference of the earth. It is fitting, then, that the octalinn should be accepted as the measure of astronomical distances.

DENOMINATIONS.

Measures of Superficies, or Square Measure—Unit ARR.

DENOMINATIONS.	VALUE ARR.	AMERICAN AND ENGLISH EQUIVALENTS.	FRENCH.
Decarr . . .	10,000,000,000	38.6102399 square miles.	
Octarr. . . .	100,000,000	0.3861 sq. mile = 247.1 acres.	
Hexarr . . .	1000,000	2.47105535364 acres.	hectare.
Tetrarr . . .	10,000	0.0247 acre = 119.519 sq. yards.	ARE.
Duarr	100	1.195990791164 square yards.	centiare.
ARR	1	0.10763917120474 square feet.	
Biniarr . . .	$\frac{1}{100}$	0.1550004065348281 square inch.	
Quartiarr .	$\frac{1}{10000}$	0.001550004 sq. inch = sq. binilinn.	

100	quartiarr	(qA)	make	1	biniarr
100	biniarr	(bA)	"	1	ARR
100	ARR	(A)	"	1	duarr
100	duarr	(dA)	"	1	tetrarr
100	tetrarr	(tA)	"	1	hexarr
100	hexarr	(hA)	"	1	octarr
100	octarr	(oA)	"	1	decarr

The decarr will serve to measure the area of continents, States, provinces, and all very large tracts of territory, more conveniently than the irregular vague quantities known as "square leagues," and "square miles," of which nearly all countries have their own, differing from others. The decarr is one square pentalinn. For other less extensive yet large bodies of land, the octarr—one square tetralinn—may be employed.

The hexarr, tetrarr, and duarr—equivalents of the French

hectare, are, and centiare—would be used, as the acre with us, for measuring smaller bodies of land, farms, lots, etc.

The French "Metrical System" is defective, and unfit for adoption as the universal system, in that it makes no provision for the measurement of area or surface, over the hexarr, nor under the duarr. It simply provides very limited agrarian measure; whereas the Linn-base System provides for the measure of all area, from that of a pin's head to a continent, the decarr containing thirty-eight and a half square miles, and the quartiarr, one square binilinn.

The decarr is the equivalent of 1 square *myriamètre* of the French system; and the quartiarr, of 1 square *millimètre.*

DENOMINATIONS.

Measures of Volume, or of Solids—Unit SOLL.

DENOMINATIONS.	VALUE SOLL.	ENGLISH AND AMERICAN EQUIVALENTS.	FRENCH.
Octasoll....	100,000,000	130,795.184 cubic yards.	
Hexasoll...	1000,000	1307.95 " "	
Tetrasoll...	10,000	2.759 cords = 13.0795 cubic yards.	décastère.
Duasoll....	100	3.5314699712460514851 cubic feet.	décistère.
Soll	1	61.0238011031317696635 cubic inches.	
Binisoll....	$\frac{1}{100}$	0.61 cubic inch.	
Quartisoll..	$\frac{1}{10000}$	0.0061 cubic inch.	
Senisoll....	$\frac{1}{1000000}$	0.000061 cubic inch = cubic binilinn.	

100	senisoll	(sS)	make	1	quartisoll
100	quartisoll	(qS)	"	1	binisoll
100	binisoll	(bS)	"	1	Soll
100	Soll	(S)	"	1	duasoll
100	duasoll	(dS)	"	1	tetrasoll
100	tetrasoll	(tS)	"	1	hexasoll
100	hexasoll	(hS)	"	1	octasoll (oS)

The department of the French "Metrical System" corresponding with this, is, like the last, (square measure), very shortcoming. It is not worthy to be accepted in a general system, as affording a satisfactory measure of solids. It affords simply—and was so intended—a measure of firewood. The Linn-base Decimal System, in its department of solid measures, completely supplies the desideratum; affording denominations which cover all bulks, from the cubic binilinn to 100,000,000 cubic linn—from sixty-one millionth parts of a cubic inch (cubic *millimètre*), to hundreds of thousands cubic yards.

In fact, the French system lacks the power to express our highest and lowest denominations, except by terms of impossible acceptance. For instance: Its equivalent of our quartiarr would be *centi-milli-milliare;* and of our decarr, *hecto-myriare.* Our senisoll would be found in *milli-milli-millistère;* and our octasoll, in *déca-myriastère.* Our octalinn would be recognized in *kilo-myriamètrè.* It is thus demonstrated that the French nomenclature will not suffice for a complete metrical system. Indeed, it confesses its own insufficiency, and gives up in despair, by retaining the old style of "*tonneau*" rather than accept the systematic denomination of *hecto-myriagramme;* and by calling its weight of 100 kilogrammes, *Quintal Métrique*, instead of *déca-myriagramme.*

DENOMINATIONS.

MEASURES OF CAPACITY—LIQUID AND DRY MEASURE—UNIT CAPP.

DENOMINATIONS.	VALUE CAPP.	BRITISH AND AMERICAN EQUIVALENTS.	FRENCH.
Triacapp..	1000	220.08 gal. = 27.51 bushels = 1.048 tuns.	kilolitre.
Duacapp..	100	2.751 bushels = 22.008 imperial gallons.	hectolitre.
CAPP	1	1.76068 pints = 0.88 quart.	LITRE.
Binicapp..	$\frac{1}{100}$	0.0176068 pint.	centilitre.
Ternicapp.	$\frac{1}{1000}$	0.00176 pint = 1 cubic primilinn.	
Senicapp..	$\frac{1}{1000000}$	0.00000176 pint.	

1000	senicapp	(sC)	make	1	ternicapp	
10	ternicapp	(tnC)	"	1	binicapp	
100	binicapp	(bC)	"	1	CAPP	
100	CAPP	(C)	"	1	duacapp	
10	duacapp	(dC)	"	1	triacapp	(tC)

Liquid and Dry Measures of capacity in common use, would be as below:

MEASURES OF	AMERICAN AND ENGLISH EQUIVALENTS.	FRENCH.
100 CAPP	22.008 gallons = 2.751 bushels	1 hectolitre.
50 "	11.004 " = 1.3755 "	5 décalitres.
20 "	4.4016 " = 0.54 " = 2.2 pecks.	2 "
10 "	8.80339 quarts=1.10 pecks=17.6 pints.	1 "
5 "	4.4 " =0.55 " =8.8 "	5 LITRES.
2 "	1.760678638 " =3.52136 "	2 "
1 "	0.880339319 " =1.76068 "	1 "
50 binicapp	0.44 " =0.88 "	5 décilitres.
25 "	0.44 pints	2.5 "
10 "	0.176 "	1 "
5 "	0.088 "	5 centilitres.
2 "	0.0352 "	2 "
1 "	0.0176 "	1 "
5 ternicapp	0.0088 "	5 millilitres.
2 "	0.00352 "	2 "
1 "	0.00176 "	1 "

1 . 25 . 50 . 75 . 100 . 250 . 500 . 750 senicapp.

Measures of 10, 20, 50, and 100 capp, made of wood, are used in French dry measure, as we use the bushel, half-bushel, peck, etc.

A metrical system which aims at completeness, and seeks universal adoption—which would satisfy all needs, and meet all reasonable requirements—which descends, in its measures of area, to the minute quartiarr, and of volume, to the almost atomic senisoll, must not fail to commence its series of practical measures of capacity, by vessels of which the content shall be of one cubic binilinn and one cubic primilinn. These are quite appreciable and measurable quantities. It is time to have done, in scientific parlance, with those inexact, uncertain expressions, so common in pharmacy, and in the practice of medicine, teaspoonful, table-spoonful, and drops, unless indeed—which is much to be desired—the several sizes of spoons were modeled, by common consent of manufacturers, upon well-known systematic quantities. This idea would, probably, soon be realized, if a metrical system providing such quantities, were generally adopted by the nations.

It is one of the characteristics of the Linn-base Decimal System, recommending it particularly to the acceptance of scientific men, and to general adoption, that the several parts bear a far more direct and simple and scientific relation to the base, and to each other, than obtains in the "Metrical System" of France. Each unit being immediately derived from the base-linn itself, and not from a multiple, or from a fraction of it—the soll and capp and poudd being cubes of the linn, and the arr, its square,—there results, for the whole, a seemly and practically useful harmony, which is wanting in the metrical system. In the last-named system, the *are* is the square of a *décamètre;* the *litre* is the cube of a *décimètre;* the *gramme* is the cube of a *centimètre*, and the *stère* alone, comes immediately from the *mètre* itself, being its cube. From this variety springs a confusion, in reference to the *Mètre*, of the numerals employed to form the several denominations,—a confusion which is distracting to most minds, and which mars the beauty of the system, as a whole. The harmony referred to, and the parallelism of the several departments, and the absence of those qualities in the French system, are made apparent to the eye, as well as to the mind, by inspection of the Synoptical Comparative Table hereto annexed. In that Table is exhibited the relation of all the parts to the base-linn, and to each other; and then, the *mètre*, with the "Metrical System" beside it, being placed in juxtaposition, and in proper relation to the Linn, the relation of the several parts of the Metrical System to its base-mètre is seen, and the two systems compared.

This harmony has been spoken of as not only seemly, but practically useful. In proof, be it remarked, that it puts to every man's hand, and in the possession of every family, for domestic purposes, a ready, convenient, and perfectly simple substitute for any of the regular systematic weights or measures—a substitute, not of scientific exactness, but very nearly equivalent, sufficiently so, for all ordinary purposes. In a 1 capp measure of common water, we have the equivalent of a 1 pondd weight. In 3, 4, 15, 50, 100 capp of water, we have the equivalents of 3, 4, 15, 50, 100 pondd of weight. One binicapp of water is equivalent to 1 binipondd of weight. And so of all the systematic measures of capacity, from the 1 binicapp to the triacapp—from one third of an ounce to 2200 pounds. *Any of the measures, filled with water, supplies the weight of its numerically corresponding pondd.* Indeed, it will be only necessary to know the contents in capp, of any vase, can, barrel, hogshead, or reservoir containing liquids, to tell instantly, with fair approach to accuracy, its actual weight in pondd, by the simple mental substitution of "pondd" for "capp." And *vice versâ*, if a capp or binicapp-measure be wanting, weigh in any other vessel, a pondd, or a binipondd of water, and you have the content desired.

It is urged against a decimal system of measures and weights, that it "has been found, in practice, unsuited to the purposes of retail traffic, to which, in fact, only a binary system, or the division of the unit into halves and quarters seems applicable." Now, it is asked, in all seriousness, if with any candor, or justice, or common sense, such objection can lie against the Linn-Base Decimal System here described? All of its units—measures and weights to be upon all counters, and daily used in "retail traffic," *are divided into hundredths.* Is it not just as easy, we demand, and just as convenient to ask for, to give, and to receive *seventy-five primilinn* of calico, as "three quarters of a henalinn" of that fabric? Is it not just as easy and convenient to ask for, to give, and to receive *twenty-five binipondd* of sugar, as "a quarter of a pondd" of the article? or *fifty binicapp* of milk, as "half a capp?" Why, these terms, for all possible "purposes of retail traffic," practical and theoretical, are respectively equivalent.

These objectors to decimal systems for "retail traffic" forget, or are not aware of this mode, (by division of units into hundredths), of rendering decimal systems "binary."

DENOMINATIONS.

Measures of Gravity—Weights—Unit PONDD.

DENOMINATIONS.	VALUE PONDD.	ENGLISH AND AMERICAN EQUIVALENTS.	FRENCH.
Triapondd..	1000	2204.8571428 lbs. av.d.p.=0.984 ton.	Tonneau.
Duapondd..	100	1.9686 cwt.=220.48 lbs. avoirdupois.	Quintal Métrique.
Pondd.....	1	15434 troy grs.=2.679513888 troy lbs.	kilogramme.
Binipondd..	$\frac{1}{100}$	0.321541 troy oz.=0.352777 oz. av.d.p.	décagramme.
Quartipondd	$\frac{1}{10000}$	0.07717 scruple=1.5434 grains.	décigramme.
Senipondd..	$\frac{1}{1000000}$	0.015434 grains=1 cubic binilinn.	milligramme.

100	senipondd	(sP)	make	1	quartipondd
100	quartipondd	(qP)	"	1	binipondd
100	binipondd	(bP)	"	1	Pondd
100	Pondd	(P)	"	1	duapondd
10	duapondd	(dP)	"	1	triapondd (tP)

This system employs but four denominations of practical weights, as seen in the following category; yet it affords greater facilities, and is no less comprehensive, than the "Metrical System," which employs eight denominations, nor than the English and American systems, which, with their troy, avoirdupois, and apothecaries' weights, count thirteen denominations.

The Weights, to be used under this system, are as follows:

Number and Denominations of Weights used in the Linn-Base Decimal System.

No.	Denominations.	American and British Equivalents.	French.
one	of 75 pondd	200.9635 lbs. troy=165.364275 lbs. av.d.p. ..	7.5 myriagrammes.
one	" 50 "	133.97569 lbs. troy=110.24285 lbs. av.d.p....	5 "
one	" 20 "	53.59 lbs. troy	2 "
two	" 10 "	22.048571 lbs. avoirdupois..................	1 "
one	" 5 "	11.024 lbs. av.d.p. = 13.4 lbs. troy.......	5 kilogrammes.
two	" 2 "	4.4 lbs. " = 5.359 lbs. "	2 "
two	" 1 "	2.2 lbs. " = 2.6795 lbs. "	1 "
one	" 75 binipondd	1.65 lbs. " = 2.0096 lbs. "	7.5 hectogrammes.
one	" 50 "	1.1 lbs. " = 1.33975 lbs. "	5 "
one	" 20 "	0.44 lbs. " = 0.5359 lbs. "	2 "
two	" 10 "	0.22 lbs. " = 0.26795 lbs. "	1 "
one	" 5 "	0.11 lbs. " = 1.6077 ounce troy	5 décagrammes.
two	" 2 "	308.68 grains = 0.643 ounce troy	2 "
two	" 1 "	2.57 drachms = 154.34 grains..........	1 "
one	" 75 quartipondd	115.755 grains = 1.929 drachms..... ...	7.5 Grammes.
one	" 50 "	77.17 " = 1.286 drachms.........	5 "
one	" 20 "	30.868 " = 1.5434 scruples........	2 "
two	" 10 "	15.434 " = 0.7717 "	1 "
one	" 5 "	7.717 " = 0.38585 "	5 décigrammes.
two	" 2 "	3.0868 " = 0.15434 "	2 "
two	" 1 "	1.5434 " = 0.07717 "	1 "
one	" 75 senipondd	1.15755 " = ...	7.5 centigrammes.
one	" 50 "	0.7717 " = 0.03858 "	5 "
one	" 20 "	0.30868 " = ...	2 "
two	" 10 "	0.15434 " = 0.007717 "	1 "
one	" 5 "	0.07717 "	5 milligrammes.
two	" 2 "	0.0308 "	2 "
two	" 1 "	0.015434 "	1 "

All the above weights have their exact equivalents in the usual French system, except the 75^{s}. But these last will be found convenient, and almost necessary for ready combination in the higher tens.

MONEY.

UNIT—MONN.

DENOMINATIONS.	VALUE.	AMERICAN.	BRITISH.	FRENCH.
	Monn.	*$. c. m.*	*£ s. d.*	*rFs c.*
Henamonn....	10	9.35.	2.0.4.5	50.00.
Monn.........	1	0.93.5	4.0.45	5.00.
Centt.........	$\frac{1}{100}$	0.9.35	0.4845	05.
Mill	$\frac{1}{1000}$	0.935	.04845	0.5

---

10 mill (m)	make	1	centt	(c)
100 centt	"	1	MONN	(M)
10 MONN	"	1	Henamonn	(hM)

COINS TO BE USED UNDER THIS SYSTEM.

DENOMINATIONS.	VALUE.			
Gold.	*M. c. m.*	*$ c. m.*	*£ s. d.*	*Frs. c.*
Double-Henamonn..	20.00.0	18.70.0	4.0.9.	100.00
Henamonn	10.	9.35.	2.4.5.	50.
Demi-Henamonn....	5.	4.67.5	1.0.2.25	25.
Silver.				
Monn..............	1.	93.5	.4.0.45	5.
50 centt	.50.	46.7.5	.2.0.225	2.50
25 centt	.25.	23.3.75	.1.0.1125	1.25
10 centt	.10.	9.3.50	4.845	.50
5 centt............	.05.	4.6.75	2.4225	.25
Copper.				
2 centt	.02.	1.8.7	.969	.10
1 centt............	.01.	9.350	.4845	.05
5 mill.............	0.5	4.675	.24225	.02.5

In order to familiarize the people promptly as possible with the new system, and for their real convenience too, in that it would enable them, in emergencies, to establish and verify its weights and measures, the diameter and weight of coins—particularly of the silver and copper coins—should be stamped upon one of their faces; and a graduated line, divided into binilinn, drawn through the centre. By these means, in con-

2

nection with the use of the capp and binicapp, as above described—and when the reformation of spoons, as previously suggested, shall be effected—it would not be long after the adoption of the system, ere every family in Christendom, without any special effort to that end, and without a teacher, might become perfectly acquainted with the theory of the system, and able to extemporize for prompt daily use, in town and country, in hut and palace, in the workshop, in the kitchen, in the parlor, its three practical units, the linn, capp, and pondd, with their multiples and fractions. Indeed, a single centt, if stamped as recommended, could, in case of need, supply all of three units. Its diameter would establish the linn. Its weight would give the pondd. The pondd and a mug of water would establish the capp.

It will be observed that the Linn-Base system adopts the French money, without other change than that of name, and the substitution of the five-franc piece, instead of the franc, as the monetary unit. This is done, not because the sum of five francs is deemed positively the best, and most suitable value for a monetary unit; nor because it is believed to possess any advantage over the American dollar, entitling it to the post of honor. It has no such advantage. Indeed, the dollar is preferable, in that it is slightly superior in value; and possessing, equally with the French coin, the decimal character, it has, moreover, the incontestible and very important advantage of being already universally and favorably known in the world's commerce, as the monetary unit of one of the leading commercial powers. Its adoption, therefore, would subject commerce and peoples to comparatively little disturbance and difficulty. But the concurrence of France in the projected reformation is very desirable. The painful throes of recently-accomplished reform within her own limits, are but just subsiding; and she would very naturally recoil from their renewal soon, or at all. Her "Metrical System" is admirable and glorious. The world is daily according to her the praise due to its achievement. It constitutes the first great step in the path of much-needed reform; and will never be forgotten by an admiring and grateful world. But the system is imperfect in its very design; and is incapable of being made suitable for adoption by the nations, as the one, common, satisfactory system. Its monetary department, if not positively the best that could be devised, is very good; and, with the changes proposed, might become acceptable in a general system of measures, weights, and money. It has, therefore, been adopted and incorporated into the Linn-Base System in the hope, if the Linn-Base System should find favor with the world, and be deemed suitable for general acceptance, as the common system, that France may be induced, by this concession in her favor, to join the family of nations in its adoption; and again to meet, for her own ultimate good, and the convenience of the world, the just quieted troubles incident to such change.

But if—other nations being disposed to accept the new system—France should positively refuse her concurrence, then, it would be advisable, in order to disturb the world's commerce, and the interior life of nations, as little as possible, to reject the five-franc piece, and assume the American dollar as the monetary unit. Or, let the nations, persevere in the efforts now being made to find some other more convenient and acceptable unit of money than the American dollar; and applying to it, when found, the names and other suggestions herein proposed, incorporate it into the Linn-Base System of measures and weights.

A word, in conclusion, touching the exact scientific length of the standard-base of this new system. It is now well known, and generally admitted, that the distance of *ten million mètres,* found by the trigonometrical measurement of 1806–7 as the distance from the equator to the pole, is a little too short. Bissel found it too short by 935 yards; Puissant, by 1411 yards; Chazallon, by 1958 yards. The great surveys in India (1832–42) also present slightly varying results. If any system of measures and weights, founded upon the measurement of an arc of the earth's meridian, be adopted by the nations, the error above signalized, whatever it be, should be ascertained as exactly as possible, and formally corrected. It is certain, in advance, that the error would amount to nothing in its application to the measures and weights of commerce. The difference to be ascertained is so slight as to be quite inappreciable as affecting the length of the Linn. But for theoretic and scientific uses—astronomical and other—to say nothing of the honor of science—the length of the Octalinn should be determined with the utmost precision to which science can attain.

SYNOPSIS

OF THE LINN-BASE DECIMAL SYSTEM.

UNIT OF LINEAL MEASURE.

LINN.

(The 100,000,000th part of the quadrant of the earth's meridian.)

Measures.		*Units.*		*Value.*
Length	unit	Linn	=	3.93700910 inches of imperial yard.
Area	"	Arr	=	1 square linn = 15.5000406535 square inches.
Solids	"	Soll	=	1 cubic linn = 61.023801103 cubic inches.
Capacity	"	Capp	=	1 cubic linn = 1.7606 pints imp. gallon.
Weight	"	Pondd	=	1 cubic linn = 2.6795 troy pounds.
Money	"	Monn	=	5 francs.

LENGTH.	AREA.	SOLIDS.	CAPACITY.	WEIGHT.	MONEY.
	Decarr				Decamonn
Octalinn	Octarr	Octasoll			
	Hexarr	Hexasoll			Hexamonn
Pentalinn					
Tetralinn	Tetrarr	Tetrasoll			
			Triacapp	Triapondd	
Dualinn	Duarr	Duasoll	Duacapp	Duapondd	
Henalinn					Henamonn
LINN	ARR	SOLL	CAPP	PONDD	MONN
Primilinn					
Binilinn	Biniarr	Binisoll	Binicapp	Binipondd	centt
					mill
	Quartiarr	Quartisoll	Quarticapp	Quartipondd	
		Senisoll	Senicapp	Senipondd	

It would perhaps be well to suppress the *ternicapp*, and add the *quarticapp*; making the denominations of the Capp and Pondd correspond throughout, as in the above synopsis.

www.ingramcontent.com/pod-product-compliance
Lightning Source LLC
LaVergne TN
LVHW011146110826
845150LV00008B/2533

* 9 7 8 1 4 1 8 1 9 2 1 1 2 *